John E. Owens

Sketch of the Medical Bureau

World's Columbian Exposition

John E. Owens

Sketch of the Medical Bureau
World's Columbian Exposition

ISBN/EAN: 9783337801311

Printed in Europe, USA, Canada, Australia, Japan

Cover: Foto ©berggeist007 / pixelio.de

More available books at **www.hansebooks.com**

SKETCH OF THE MEDICAL BUREAU, WORLD'S COLUMBIAN EXPOSITION.

BY JOHN E. OWENS, M.D.

MEDICAL DIRECTOR OF THE MEDICAL BUREAU OF THE
WORLD'S COLUMBIAN EXPOSITION.

CHICAGO, ILL.

REPRINTED FROM
THE JOURNAL OF THE AMERICAN MEDICAL ASSOCIATION.
MARCH 16, 1895.

CHICAGO:
AMERICAN MEDICAL ASSOCIATION PRESS.
1895.

SKETCH OF THE MEDICAL BUREAU WORLD'S COLUMBIAN EXPOSITION.

BY JOHN E. OWENS, M.D.

ROSTER.

Medical Director.--Dr. J. E. Owens, June 1, 1891, to March 31, 1894.

Attending Physicians.—Drs. N. R. Yeager, June 1, 1891, to Nov. 30, 1893; W. H Allport, June 1, 1891, to Nov. 30, 1893; S. C. Plummer, August 1, 1891, to Nov. 30, 1893; G. P. Marquis, Jan. 1, 1892, to March 1, 1893; E. T. Edgerly, Feb. 1, 1893, to Nov. 15, 1893.

Resident Physicians.—Drs. J. L. Hillmantel, April 10, to Nov. 30, 1893; D. M. Appel, Captain and Ass't. Surgeon, U. S. A., May 1, to May 17, 1893; W. C. Raughley, May 17, to Oct. 30, 1893.

Hospital Steward.—William Kelly, April 7, to Dec. 20, 1893.

Druggists.—Harry Kahn, April 16, to Sept. 1, 1893; F. C. Cady, Sept. Sept. 1, to Nov. 16, 1893.

Superintendent of Nurses.—Mary R. Browne, March 16, to Nov. 30, 1893, (St. Luke's Training School, Chicago).

TRAINED NURSES.

Harriet Fulmer, Anne Northwood, March 15, to April 15, St. Luke's, Chicago.

Karin Eckstrom, March 15, to November 15, Stockholm, Sweden.

Helen Parr, April 16, to August 1, Louise Richey, August 1 to December 1, St. Luke's Chicago. (Surgical Nurses).

Bessie Woolings, May 1, to June 1, Guy's Hospital, London, Eng.

Emma Dawson, May 1, to June 15, St. Luke's, Chicago.

M. McDonald, E. McDonald, May 15, to June 15, Boston City Hospital.

M. Clark, A. Smith, May 15, to June 15, Massachusetts General Hospital.

Mary Forbes, June 1, to August 15, St. Luke's, Chicago.

Muriel Moberly, different terms between June 15, to November 15, St. Luke's, Chicago.

Helen Wiltsie, June 15, to July 15, St. Luke's, Chicago.

Susan Read, Mary Townsend, June 15, to July 15, Johns Hopkins Hospital.

Catherine DeWitt, Eliza Moore, June 15, to July 15. Illinois Training School.

Clara E. Parsons, Marie A. Lawson, July 15, to August 15, Michael Reese Hospital.

Anna Bartle, July 15, August 15, Toronto General Hospital.

Alice Cooper, Lockhart Baillie, July 15, to August 15, Bellevue Hospital.

Bessie Wolfendin, August 15, to September 15, Toronto General Hospital, Toronto, Canada.

Alice Buckman, M. L. Van Thyne, August 15, to September 15, Philadelphia Hospital.

I. C. Mathis, A. N. Bartholomew, August 15, to September 15, University of Pennsylvania Hospital.

Ruth Williams, Helen Barnard, September 15, to October 15, Johns Hopkins Hospital.

Martha Munn, Margaret Munn, September 15, to October 15, New York Hospital.

Virginia Williams, August 15, to October 1, St. Luke's, Chicago.

Mary Shears, July 15, to October 1, St. Luke's, Chicago.

Agnes McCoy, July 15, to September 15, St. Luke's, Chicago.

Gertrude Phillpotts, September 15, to November 15, St. Luke's, Chicago.

Pauline Hollenbeck, Florence Bell, October 1, to November 15, St. Luke's, Chicago.
Elizabeth Pope, October 1, to November 1, St. Luke's, Chicago.
Margaret M. Watson, Annie Louise Higgins, October 15, to November 6, St. Luke's, New York.
Nettie West, October 15, to October 30, Michael Reese Hospital.
Superintendent of Ambulance Service.—H. W. Gentles, M.D., April 10* to November 30, 1893.
Superintendent of Ambulance Corps.—J. F. Minot, April 10, to November 12. *Ambulance Corps.*—C. L. Hammond, W. K. Hammond, H. A. Adams, John T. Hutcheroft, C. Brooker, J. S. Brown, Edgar Thomas Lucas, M. E. Wheelan, J. H. Missey, A. M. Schabad, J. M. Smith, L. M. Dunavan, J. C. Egan, E. B. Hutchison, H. G. Graham, H. B. Bartholomew, W. B. Robinson, C. W. Culp, Fred Herr, J. L. Allen, T. H. Little, M. G. Bryan, Morgan Savidge, J. W. Walker, C. E. Downey, L. W. Dunavan, and others.
Ambulance drivers.—M. J. Sanborn, F. H. Crein, E. J. Pettinger, D. M. Good, J. H. McLernon, J. W. Tarbell, J. E. Price, and others.
Orderlies.—Claus Williams, A. H. Kennedy, Chas. Webster, N. K. Derderian, J. E. Johnson.
Chief Sanitary Officer.—D. M. Appel, M.D., May 17, to July 26, 1893.
Sanitary Inspectors.—J. H. Kellogg, Mary Glennon, Anna Byford Leonard.
Recording Clerk.—Alexander Goldstein, May 8 to first week of December.
Stenographer and Typewriter.—Anna H. Newton.
Messenger.—Frank Frisble.
Telephone operator.—Robert Rozelne.
Janitors.—A. U. Brown, George Blakey, E. Fleming, and others.
Ward maids.—Christine Webster, Anne Boettner, Rachel Ellingston.

DEDICATION DAY AND THE DAY FOLLOWING.

Drs. E. T. Edgerly, Jennie Hayner, Bertha Van Housen.
Trained Nurses for the Hospital (Dedication Days).—E. Farrow, M. Moberly, E. Dawson, E. Jackson, C. Fowler. *Ward maid.*—Rachel McVicker.
Trained Nurses at the two Substations in Manufactures Building.—M. R. Browne, Louise Richey, M Bryce, R. A. Ellerby.
Orderly at Hospital.—H. T. Haas.
Orderlies at Substation.—J. W. Walker, O. O. Witherbee.
Ambulance Surgeon.—Dr. G. P. Marquis.
Litter Bearers on Ambulance.—J. A. Pohling, C. Ehling.

ORGANIZATION.

The last week in May, 1891, the Committee on Grounds and Buildings established the Medical Bureau, and, after a conference with Mr. E. T. Jeffery, Chairman of said Committee, the position of chief medical officer of the Exposition Company was accepted. On May 29 the following general arrangement was agreed upon and approved:

That the title of the chief medical officer should be "Medical Director;" that the appointment be effective June 1, 1891; that the Medical Director shall have authority to appoint, subject to the approval of the Committee on Grounds and Buildings, such assistants as he may deem necessary for efficiency in organization and the prosecution of the work.

The duties of the Medical Director were thus formulated: To organize a medical bureau having jurisdiction over all cases of personal injury occurring in the line of duty; to assume the care of the resident population and visitors, and such other cases as the Exposition Company might be responsible for, or in its judgment should be cared for; to exercise a supervision over all sanitary and hygienic matters connected with the grounds and buildings, and over all matters pertaining to the health and physical well-being of the construction employees engaged in the conduct of the Exposition.

In the pursuance of some of the foregoing objects, it was decided to establish a temporary hospital upon the grounds and afterward, as circumstances might require, and after conference with the Committee on Grounds and Buildings and the Construction Department, to construct a building better suited to meet the exigencies of the Medical Bureau.

The Medical Director was instructed to report to and receive his instructions from the Committee on Grounds and Buildings, and to confer from time to time, as circumstances might suggest, with Mr. D. H. Burnham, the Chief of Construction, or his representatives, in order that coöperation might be secured in accomplishing the purposes in view.

It was agreed that the Medical Director should keep such reports and data regarding his bureau as would enable him to make-monthly reports to the Committee on Grounds and Buildings, and a final report at the close of the Exposition.

Blank pay rolls, vouchers and other stationery required were furnished on the order of the Medical Director upon the Grounds and Buildings Committee.

A monthly pay roll for himself and assistants was made by the Medical Director, and the same transmitted to the Committee on Grounds and Buildings; and, at the same time, duly certified bills and vouchers for expenses. The attending physicians treated all cases of illness and injury brought to the emer-

gency hospital, and devoted more or less time to
sanitary inspection during the construction period.
The grounds were districted in order to facilitate
this work.

The large resident population, together with the
great number of workmen about the grounds at all
times, necessitated the appointment of resident physicians who lived in the emergency hospital. As
these gentlemen were required to reside in the hospital they could not, therefore, engage in private prac-
tice. The duties pertaining to the office of resident
physician consisted for a time in the inspection of
grounds and buildings; care of the resident population; they were responsible for the efficiency of the
hospital service; for the conduct of the hospital in
general during the absence of the Medical Director;
they furnished a report of deaths, weekly and daily
reports of the work performed and, as occasion required, they aided the attending physicians in the
performance of their work.

It gives me pleasure to testify to the ability with
which both attending and resident physicians performed their onerous duties; to their loyalty to the
service; to the promptitude with which they filled
their daily appointments and to the courtesy and
helpfulness extended to the sick and injured. All of
this largely contributed to the success of this department of the Exposition.

On July 15, 1892, the Medical Bureau was placed
under the auspices of Mr. D. H. Burnham, Chief of
Construction.

BUREAU BADGE.

Upon the flag of this department is seen, in the
center, the Geneva cross and the words, "Medical
Bureau." The badge worn by the physicians consists
of the Geneva cross, containing in its center a disc, in
which is a countersunk panel showing in raised design
the lettering, "World's Columbian Exposition,"
"Medical Bureau," and in the center is the symbol
of Hygeia. Interwoven and forming a wreath inside

of the points of the cross are, in raised relief, poppy heads, signifying ease and comfort. The central circular portion of the badge and poppies are silvered. The badge of the Medical Director has a gold center, the circumference of which is of blue enamel, and the poppies silvered. This, with the sealing wax red of the Geneva cross, presents the national colors—red, white and blue. An illustration of the Medical Director's badge may be seen at the head of the official letter paper; and also upon the sides of the ambulance; on the doctor's buggy, and over the east entrance to the hospital. The nurses' badge is that of the attending physicians, minus the poppy heads. All other attachés of the Medical Bureau wore the central circular portion of the badge, silvered, without cross and without poppies.

BUILDINGS.

For the beginning of the dispensary work, temporary quarters were at once secured in a corner of a room then used by the Construction Department. Soon after, a one-story building was erected, consisting of a waiting room, a sleeping room, an operating room and a store room. This was speedily equipped and occupied July 20, 1891. The hospital was soon after erected, and formed a part of the Service Buildings at Sixty-second Street. Sept. 3, 1892, we moved into this building, but it was at that time mostly occupied by a portion of the clerical force of the Construction Department. Work began here in four rooms. Oct. 19, 1892, the south wards were made available—one for men and the other for women. Feb. 1, 1893, the operating room was secured. April 24 to 28, 1893, the north wards were evacuated by the clerical forces, and still later, the room occupied by the telegraph operators.

The emergency hospital contained on the first floor the following: Two wards of ten beds each for men; one ward of ten and another of three beds for women; Medical Director's office; resident physicians' office;

attending physicians' office; office of the Superintendent of Nurses; a diet kitchen; drug room; linen and clothes closet; two examining and operating rooms and closets; reception room and two waiting rooms—one for men and one for women. In addition to the beds above noted, ten rattan lounges were distributed about the wards and corridors. On the second floor were the dormitories for nurses, resident physicians and druggist. No meals were served in the hospital for either patients or staff, but in the diet kitchen the nurses prepared soups, teas and refreshing and stimulating delicacies. It was a rare exception to keep a patient over night.

In cases where transportation of patients to the hospital is rendered difficult or impossible in consequence of dense crowds, temporary substations or portable hospitals should be established. From five to ten beds in each, according to circumstances, would be sufficient in most cases. They answer admirably, provided precautions be taken to prevent the demand from over-reaching the capacity. On such occasions even the passage of carriages through the street on their way to the hospital for the purpose of conveying patients to their homes, railway stations, or hotels, was extremely difficult and time-consuming.

On Chicago Day when there were 716,881 paid admissions, the Exposition being open both day and night, the necessity for these extra provisions was very great.

TRAINED NURSES.

The desirability of employing only trained nurses for the emergency hospital was apparent from the outset, but the expenses pertaining thereto would have been considerable at the ruling rates in this city. The scheme finally adopted was as follows: To offer an opportunity to the leading training schools in the country for the representation of their schools by the sending of two trained nurses to serve thirty days at $25 a month, with allowances, such as board, lodging and laundry service. This offer was readily accepted. The nursing force was placed in the charge of Miss M. R. Browne, a graduate of St. Luke's Training School for Nurses, Chicago. Her title was "Superintendent of Nurses." Her appointment was made effective March 15, 1893. Exclusive of the superintendent, at first there were ten nurses, one of whom was called "the surgical nurse." The latter had charge of the operating room and the preparation of instruments for operations, and when on duty, was always in service at the operating table. Each nurse was on duty eight hours and off sixteen hours. In this way they were able, while giving their schools representation in the work, to study the Exposition and at the same time render us that skilled service which is now so much in demand in our own and other countries. They remained in service from 8 A.M. until 11 o'clock at night. After that hour a sufficient force was always available at the hospital at a moment's notice, however great the demand. The following institutions were represented: Bellevue Hospital, New York Hospital, St. Luke's Hospital, New York; Philadelphia Hospital, Hospital of the University of Pennsylvania, Johns Hopkins Hospital; St. Luke's Hospital, Chicago; Michael Reese Hospital, Illinois Training School, Massachusetts General Hospital, Boston City Hospital; Guy's Hospital, London, England; one in Stockholm, Sweden, and Toronto General, Canada. We had a greater number from St. Luke's Training School, Chicago, than from any other, and made a point of keeping a nucleus of nurses from that institution, familiarized with the work. After the first month, in addition to the Superintendent of Nurses, we never exceeded twelve on duty at one time, except during three or four weeks subsequent to the Cold Storage disaster, when fourteen nurses were required. By common consent four nurses were known as "the pioneer nurses," namely Miss M. R. Browne, Superintendent, and the Misses Fulmer, Eckstrom and Northwood. These, under the guidance of their able superintendent, in a building at that time a notable thoroughfare and well filled with officers and clerks whom we often despaired of dis· lodging, in the midst of the legitimate professional work then going on, including the examination of many hundreds of guards, without exaggeration, absolutely quarried their way through many obstacles and eventually there appeared a hospital that we believe was pleasing to all.

As attachés of the nursing force there were four orderlies and two ward maids. Their duties consisted of stretcher work, lifting patients, assisting in the examining room, cleaning, etc. In addition to these, four colored janitors took care of the building and kept the drug room in order. Two telephones— one for "long distance" and one for "short distance"— were early found necessary.

When fully equipped, the personnel of the emergency hospital consisted of the following: Two resident and four attendant physicians, a superintendent of nurses, with a corps of twelve trained nurses, four orderlies, a druggist, a recording clerk, a messenger boy, a telephone operator, four janitors and a hospital steward, who had care of the property, records, reports, and who transacted the general business pertaining to the Medical Bureau. For a time daily, and on all special occasions, when the crowds were the greatest and the demands upon the hospital were proportionately increased, Columbian guards were detailed for service at the entrance of the hospital.

All were uniformed—the physicians in navy blue, the orderlies and litter bearers in gray and the nurses in the uniforms of their respective schools, except the superintendent of nurses who wore white.

A general kitchen was not included in the plan, and therefore no meals were served in the hospital. The nurses ate at the officers' mess at hours specially fixed. The diet kitchen was used solely for the preparation of special diet for the sick and injured. It was a rule of the department that patients should not be kept over night. Exceptions were made only when absolutely necessary.

AMBULANCE SERVICE.

During the pre-Exposition period the sick and injured were transported in patrol wagons. At the dedicatory and opening exercises, as well as on other occasions where the people were densely massed, invalid chairs, marked with a conspicuous red cross on white background, in charge of trained men, were stationed on the outskirts of the crowd. Sometimes the location of the chairs was indicated by means of a very large red cross on a sheet on the sides of the building and above the level of the crowd. The work of the ambulance corps proper began on May 1, with the opening of the Exposition. At first, four ambulances were put into service, but the exigencies did not seem to require the fourth one and its use was discontinued. In less than fifteen seconds after a call had been received a wagon was ready for the start. The ambulance corps numbered twenty-five men (litter bearers), exclusive of their sergeant, J. F. Minot, and nine drivers. The latter were not allowed to leave the box. Dr. H. W. Gentles, the superintendent of the ambulance service, received his instructions from the Medical Director. Dr. Gentles thoroughly drilled the ambulance corps in "first aid," and performed his duties in a very able manner. He had had, previous to his appointment, much experience in this kind of work. The greater proportion of the litter bearers were young medical men who had had more or less experience as hospital internes; some had seen service in the Army Hospital Corps; others were medical students, and all were well fitted for this special branch of the service.

Quick transportation through large crowds was a matter of extreme difficulty. This was avoided occasionally by the use of the Ashford Wheeled Litter, loaned by the St. John's Ambulance Association.

This litter and the invalid chairs above mentioned were of great service in moving patients either through large buildings or through dense crowds to points where the ambulances could continue the transportation to the emergency hospital. The principal days on which these precautions were found necessary were the Opening Day, Infanta's Day, the Fourth of July, Rajah's Day, Illinois Day and Chicago Day. The system of communication between the sick and the ambulance was as follows: A friend of the sick person called the nearest guard, who telephoned by means of the nearest patrol box his desire for an ambulance and, if possible, the nature of the case. The call on being received by the central office was transferred to the nearest ambulance station, and the ambulance wagon immediately took the shortest and most unfrequented road to the box. It was the duty of the guard to remain at the box until the ambulance wagon arrived in order to direct the bearer to the case, which was then, after the immediate indications were fulfilled, removed to the hospital. On days when there were especially large crowds, men were stationed at certain boxes marked by large red cross flags. In this way, any person who had a sick friend had no difficulty in knowing where assistance was to be received. After patients had been treated at the hospital the ambulances were frequently called into requisition, either to transfer a patient to his home, to his hotel, or, again being placed in a rattan basket stretcher, he was sent by rail to the point nearest his home or hospital, and there the city police ambulance or patrol wagon completed the transportation. The men remained on duty eight hours, there being three reliefs in the twenty-four hours. The ambulances were in service from 8 in the morning till 11 o'clock at night, and longer when necessary. One wagon, that was nearest the hospital, remained on duty twenty-four hours, so that a constant ambulance service prevailed. Two drivers were assigned to each wagon, with the exception of the hospital ambulance, which had three—two during the day and one at night. Each of two wagons had an assignment of one pair of horses. The hospital ambulance had two pairs, and in the event of any sickness among the horses, substitutes were easily obtained. The ambulance corps had several calamitous accidents to take charge of— three in particular. The fatal Cold Storage fire occurred July 10. On this occasion twenty-two cases were brought to the hospital in a very short space of time. The men also, in addition to their ordinary duty, remained at the scene of the fire and took care of any trivial accidents incident to the occasion, and also supplied coffee, sandwiches, etc., to the workers. On June 28 a floor gave way in one of the temporary buildings and injured a considerable number of people. Thirty of these were taken care of by the ambulance corps. On the occasion of the accident on the ice railway the sufferers were removed quickly and with dispatch to the hospital. On July 4, 70 calls were answered; In May, 315; in June, 406; in July, 581; in August, 564; in September, 605; on Chicago Day (October 9), 172. One ambulance was stationed at the hospital building; another contiguous to Terminal Station, Machinery Hall, and the Grand Plaza, and a third at the east end of Midway Plaisance, convenient to the north end of the grounds. While there were a number of narrow escapes, no visitor or employee was actually struck by either horse or ambulance. The latter carried, in addition to splints and litter, a bag with all necessaries likely to be called for in an emergency case. Moreover 150 light stretchers, with a single blanket for each, were enclosed in white canvas bags marked, "Stretcher" and placed in conspicuous places in the chief buildings, to be used in the event of a great accident. These were easily obtained.

RULES FOR AMBULANCE SERVICE.

1. All calls are to be taken on the telephone by No.

1, who shall repeat the message over again to the operator.

2. No. 1 shall in all cases inform the driver whether there is any special reason for haste.

3. On receipt of a call, the wagon is to leave the station in the shortest possible time. All unnecessary bustle and noise is to be avoided.

4. The driver shall always take the most unfrequented and shortest route to and from an accident. No driver shall use the gong when not necessary, but if path is crowded he is to keep gong going so that everybody has sufficient time to get out of the road. The gong must always be rung before turning a corner. All guards or guides shall do everything in their power to facilitate the progress of the wagon whether they be on duty or off.

5. The man on the wagon must always report the name and number of any guard who does not assist in clearing the way for the wagon; also any person, carriage or chair so placed, driven or wheeled as to obstruct the passage for the wagon.

6. All reports are to be handed in (in writing) as soon as possible to the Superintendent of Ambulance, so that he may investigate the report and forward the result to the Medical Director for instructions.

7. If, on arriving at the place from which the call has been sent, there is nobody at the box to direct the ambulance men this fact must be reported at once to the Superintendent.

8. All members of the corps *must* show the utmost consideration to sick and injured persons and their friends, especially ladies. They shall always try to accommodate one of the patient's friends inside the wagon; otherwise nobody is to be allowed to be inside or on the steps of the wagon.

9. Men off duty are *never* to ride on the wagon.

10. The ambulance of the Bureau, while engaged in going for or in carrying sick or wounded persons to or from the hospital or substation, shall have the

right of way, excepting fire apparatus responding to
alarms of fire, against any person, carriage or incum-
brance, put, driven or being in streets, and no person shall obstruct said ambulance while so engaged
if there shall be an opportunity to get of the way.
By the courtesy of the Chief of Police of the city
of Chicago, Major McClaughry, this rule holds good
outside the grounds, and city police officers have instructions to give ambulances of the Bureau the right
of way in the streets of the city of Chicago.
Approved.

Jno. E. Owens, Medical Director.
D. H. Burnham, Director of Works.

WATER.

The statistics of the Department of Health of the
City of Chicago showed, to February, 1892, a very
serious death rate from typhoid fever, the number
of deaths from this cause for the year ending Feb.
29, 1892, being nearly two thousand. After this date
the water supply of the city was much improved by
the extension of tunnels. The general supply of
water for the city, contiguous to the Exposition
grounds, was obtained from the lake through the
pumps at Hyde Park Station at 68th Street. The
supply to the grounds was independent of that to
the city, and furnished by means of pumps erected
at Hyde Park Station by the Exposition Company,
and to be taken by the city, after the close of the
Exposition. The amount of water which could be
pumped was 24,000,000 gallons daily, but the domes-
tic consumption on the grounds averaged about eight
million gallons per day. The extension of the tun-
nel into the lake had not been completed, and pure
water for the workmen became a desideratum. As
soon as practicable, arrangements designed and constructed by Mr. W. S. MacHarg, Engineer of Water
Supply, Sewerage and Fire Protection, were completed for the supply of sterilized water for the use
of constructionists and others. The method of pre-

paring the drinking water was as follows : Adjacent to the boiler room at the temporary power house four tanks were set, the number being afterwards increased to seven, graded and elevated so that there was a difference of level between each successively. Each tank was seven feet in diameter and nine feet high and made of cypress. The upper tank was used for cooling, and a coil of inch pipe was placed in it, delivering at the top on a pan from which the water flowed in a thin sheet Into the next tank below and overflowed through a pan in the same manner to the successive tank. The water was thus aerated and cooled at the same time. A feed water heater was set in the boiler room, and the water delivered under pressure through this heater into the coil, the steam in the heater being at ninety pounds pressure, and the water supply throttled so that a temperature greater than 212 degrees Fah. was obtained. From the lowest tank the water was drawn off at a point about two feet above the bottom into tank carts for delivery on the ground. At this time there were forty-one barrels or casks in service, with an additional thirty-five for use whenever required by any increase in the force. Each of the barrels was emptied and refilled daily, without regard to the amount of water it may have contained on the arrival of the water cart. Every Sunday the tanks were emptied and washed out thoroughly with a hose. In tank No. 1, where the boiling was done, there was a great deal of mud precipitated during the boiling of the water. There was no mud in remaining tanks. At this date (May 20, 1892), with forty-one barrels, one water cart, and with two men, the service was efficient.

Although we had been boiling the water for the use of the men on the grounds, and so far as in our power had made it convenient for them to use it, it was with more or less difficulty that the employees were prevented from using hydrant or surface water. We did not at this time consider the untreated lake water safe, and notices were distributed prohibiting the use of water from hydrants and lagoons.

In April, 1893, before the opening of the Fair, the number of barrels in use amounted to about three hundred. It was not intended to carry the use of this sterilized water into the period of the Fair, the filtered water and the Waukesha water (spring water) being intended to serve the general public, but as the • number of visitors increased in July and August it was found necessary to revert to the system in order to furnish an adequate supply. Later, the number of barrels was greatly reduced. Consequently, the sterilizing plant, which had been torn down upon the removal of the temporary power plant immediately before May 1, was re-erected in connection with the boiler plant in Machinery Hall, and this service reinstated. City water was used in the barrels previous to the re-erection of the sterilizing plant. August 3, sterilized water was again furnished, and its use continued to the end of the Fair. On certain other occasions, owing to changes in the temporary power plant, it was necessary for several days at a time to discontinue the use of sterilized water, and the barrels were filled directly from the city main. An excess of gastro-intestinal disorders was observed by this Bureau, and the Department of Water Supply, Sewerage and Fire Protection was at once notified. The sterilization of the water was of great importance, and the bacteriologic reports furnished by Mr. Allen Hazen, chemist, proved that it compared most favorably with any water in use during the Exposition.

The Waukesha Hygeia Mineral Spring Company obtained a concession for the exclusive privilege of piping water a distance of one hundred and one and a half miles and delivering it from the Springs in Southern Wisconsin, on the grounds of the Exposition, and there retailing the same from numerous booths at a cent a glass, and also furnishing it to customers at a price not to exceed five cents a gallon.

This water was passed through a cooling plant sufficient to reduce its temperature. The machinery had a capacity for reducing 60,000 gallons of water from the ordinary temperature to 40 degrees Fah. in sixteen hours. There were at one time during the summer 167 drinking booths in and outside of the buildings, and 372 private taps for wholesale delivery.

In addition to the sterilized water above referred to, free filtered drinking water was furnished at available points by means of a hundred Pasteur filters, the capacity of each being about two hundred gallons. Four faucets and four cups were furnished each filter. No attempt was made to cool this water, but its quantity and quality, as shown by the bacteriologic examinations by Mr. Hazen, were very satisfactory.

Every endeavor was made to force concessionists, State officials and restaurant people on the grounds to filter all water used for drinking. This was in a measure successful. Filters of many kinds were used by these parties, some of them probably without much benefit. This order did not apply to the Waukesha Mineral Spring Company's supply, piped from Waukesha, Wis.

So much for the efforts of the Exposition Company to furnish innocuous water to its visitors. The report of Mr. MacHarg, the Engineer of Water Supply, etc., will furnish additional details.

SEWERAGE.

The sewerage system of the city of Chicago did not extend into Jackson Park. For the care of the excreta during the construction of buildings, wooden privies with concrete floors were provided, each privy containing thirteen seats, with stalls and galvanized iron pails for solids, and troughs and sumps for fluids, and with a separate urinal trough just above the pails, and also a general urinal trough of galvanized iron along the wall facing the stalls.

An attendant threw into the pails from time to time, as became necessary, dry earth pulverized at the temporary power plant, for the absorption of dampness and the prevention of odor. A solution of copperas was from time to time freely applied to the troughs which led to the sumps, and to all the parts of the privy exposed to contamination.

A tank cart made the rounds at night, and the catch basins were pumped out and washed with copperas and the urinals carted to the nearest city sewer and dumped. The solids were conveyed to the southern portion of the grounds and dug into black earth. This proved most satisfactory except on one or two occasions of very wet weather in the spring of 1892.

Avoiding any system of contract scavengering, the Department of Water Supply, Sewerage and Fire Protection detailed its own men for the care of the privies and the disposal of the excreta. In this way the results of the system proved most satisfactory. This system was continued until the completion of the general system of sewerage and plumbing, which was shortly before the opening of the Exposition. The system last referred to, namely, for the use of the Exposition, was as follows:

All roof water was discharged directly into the lagoons or Lake Michigan, as was most convenient, through pipe sewers ordinarily, brick sewers being required in one or two cases only.

All surface water from walks, whether resulting from rain or from washing, was discharged through a system of vitrified sewer pipes to wells conveniently and economically located, and thence pumped directly to Lake Michigan. This water, while not seriously contaminated, was thought to contain sufficient organic matter to make its discharge into the lagoons undesirable. The sewage proper; that is, the discharge from water closets, lavatories, café kitchens, etc., was pumped by means of fifty-two Shone ejectors through cast iron force mains and discharged into tanks at the cleansing works, which were located in the southeastern portion of the grounds. Here sewage was treated with chemicals.

Most of the suspended matter and a large part of
the organic matter in solution were precipitated, and
a clear effluent water, from which all highly putres-
cible matter had been removed, was discharged into
the lake. The precipitant commonly called " sludge "
was forced into filter presses and pressed into the
form of cakes, which were afterwards burned at the
crematory in the extreme southeastern portion of the
grounds. These cakes came out very hard, having
been deprived of about 50 per cent. of the water by
pressure.

Garbage and ashes were handled in the following
manner: The former was placed in galvanized cans
placed outside of the buildings at 11 o'clock each
night, and taken up by the Transportation Department, Mr. W. H. Holcomb, General Manager, and
deposited at the crematory.

The ashes and garbage were kept separate. Ashes
were utilized for "filling " on the grounds. The quantity was not great, as almost all of the cooking was
done by gas.

Inasmuch as a constant fire existed at the crematory, the garbage was disposed of as soon as received.
It created no nuisance whatever, and the raw mate-
rial was very quickly reduced. The crematory was
operated by oil as a fuel, and its practicability and
efficiency were fully demonstrated.

Creolin-Pearson, in strength of one ounce to a
gallon of water, was liberally used to disinfect sta-
bles, water closets, urinals, different portions of the
grounds and buildings, and proved to be very efficient.
Tin hand sprinklers were found convenient for its
distribution.

After the opening of the Fair it was found necessary to erect a few free urinals. These were com-
posed essentially of small wooden buildings with
concrete floors and with four slate stalls emptying
through a trap directly into the sewer. They operated satisfactorily.

The whole system of sewerage was constructed and

operated by Mr. W. S. MacHarg, Engineer of Water
Supply, Sewerage and Fire Protection.

SANITARY INSPECTION.

Previous to the opening of the Exposition, the inspection of grounds and buildings was very satisfactorily performed by the attending physicians. The
grounds were divided into sanitary districts in order
the better to systematize the work. The physicians
furnished weekly reports of the condition of their
districts, covering privies and closets, water supply
and drainage, the condition of grounds, buildings
and of population.

Finally, the hospital work increased to such a
degree as to necessitate special appointments for the
performance of this function.

Inspection of grounds and buildings went on daily,
and each Inspector furnished a report to the Medical
Director every evening, covering the buildings and
localities visited, sanitary condition of the same, with
remarks and recommendations. From the reports of
the Sanitary Inspectors the Medical Director furnished each day to the Director of Works a condensed
report.

EXPENSES.

	Cr.	Dr.
Expense of Bureau to Dec. 31, 1893		$44,277.17
Expense of Bureau from Jan. 1 to March 31, 1894		638.79
Credits: Sale of bedding, furniture, instruments, phar-		
maceutical implements, ambulance	$2,788.86	
For "first aid" to injured	2,116.08	
	$4,904.44	$44,915.96
		4,904 94
Net expenditure		$40,011.02

DEDICATION.

For the exigencies of the dedicatory services on
the grounds Friday, Oct. 21, 1892, and of the dedication of certain State Buildings October 22, the following organization was effected:

1. The employment of four nurses and one head
nurse, an extra male orderly and a female orderly
for the emergency hospital.

2. The establishment of a substation (No. 1) in the

north end of the Manufactures Building (west side), divided into three rooms, one bed and a rattan lounge in each of two of these, the reception room being in the center, in charge of Dr. W. H. Allport and Dr. Bertha Van Housen, with two trained nurses and one male orderly. Another substation (No. 2), like No. 1, in charge of Dr. E. T. Edgerly and Dr. Jennie Hayner, was located in the south end of the building, east side. These were otherwise equipped for meeting such emergencies as arose. They were numbered 1 and 2, and also distinguished by the flag of the Medical Bureau. A patrol wagon inside of the Manufactures Building was at the disposal of the medical officers in charge upon application to the guard in attendance, and one of our ambulances was stationed at a convenient point.

3. Four janitors were detailed to take charge of as many wheeled chairs, designated by a red cross on white background, stationed where they would be most promptly available. Instructions were given to bring the patient to the substations by the shortest available routes. As rarely as possible the ambulance and patrol wagon were called into service for conveying patients from the substations to the hospital. Temporary provision was made for the ambulance service. Dr. G. Paul Marquis, one of the attending physicians, was detailed to accompany the ambulance and to direct and to have charge of the ambulance corps.

4. The southeast and southwest wards in the hospital were equipped, and the Medical Director's office, located between these two rooms, was furnished for the occasion.

5. Telephones were placed in the substations.

6. Certain guards were instructed to coöperate with the medical staff to preserve order, to keep clear space for the cases of accident and illness, to allow free ingress and egress of the medical officers wearing the badge of the Medical Bureau, the ambulance, the nurses in uniform and to facilitate the transfer

WORLD'S COLUMBIAN EXPOSITION MEDICAL BUREAU. SUMMARY OF WORK DONE AT EMERGENCY HOSPITAL DURING THE CONSTRUCTION AND EXPOSITION PERIODS.—CONSTRUCTION 1893.

	Construc. Period.				Exposition Period			Total.
		May.	June.	July.	Aug.	Sept.	Oct.	
New medical.	2,555	988	1,187	1,417	1,631	1,641	1,602	11,201
New surgical.	3,364	489	498	518	443	459	554	6,320
Re-treated medical.	2,738	474	538	354	406	301	383	5,189
Re-treated surgical.	3,196	670	798	978	689	586	575	7,492
	11,853	2,621	2,966	3,292	3,169	2,987	3,264	30,152
Employes, male.		2,490	2,328	2,356	2,023	1,564	1,642	12,352
Employes, female.		40	110	236	215	226	184	1,011
Visitors, male.		70	249	900	523	687	752	2,641
Visitors, female.		72	279	340	408	510	686	2,295
Total of male patients.		2,509	2,577	2,716	2,545	2,251	2,394	14,993
Total of female patients.		112	389	576	624	736	870	3,307
		2,621	2,966	3,292	3,169	2,987	3,261	18,299
Maximum daily No. of patients.		123—29th	147—17th	170—4th	140—24th	151—14th	253—9th	
Minimum daily No. of patients.		85—21st	48—4th	27—23rd	33—27th	31—2nd	33—1st	
Daily average.		81.54	98.86	106.19	102.22	90.56	108.60	
Deaths.	92	4	7	15	3	3	4	68
Births.	1 full	2 prem.	2 prem.	1 full	1 full			7
In-patients, male.		104	106	188	207	223	245	1,073
In-patients, female.		35	237	461	489	602	764	2,588
Ambulance calls.	5,870	915	406	581	564	605	790	3,261
Examination of guards.		1,072	203	73	288	317	344	7,757
Highest mean temperature.		72°—20th	80°—21st	83°—12th	82°—10th	82°—14th	70°—11th	
Lowest mean temperature.		40°—6th	58°—11th	64°—3d	58°—29th	46°—25th	33°—29th	
Total attendance.		1,531,984	3,577,834	3,977,502	4,687,708	5,808,942	7,945,480	27,529,400

of patients on wheel chairs or on foot to the substations.

7. Orders were issued to secure as direct a route
for the ambulance from the substations to the emergency hospital as the circumstances of the occasion
would permit.

8. At the expiration of the service it was the duty
of the medical officers in charge of the substations to
see that the whole medical outfit was immediately
removed to the emergency hospital by one of the
patrol wagons or the ambulance.

PATIENTS KEPT OVER NIGHT FROM MAY 1, 1893, TO OCT. 30, 1893.

	Males.	Females.	Total.
May 1, to July 31	7	3	10
August 1, to August 19	1	2	3
August 19, to August 26	5	.....	5
August 26, to September 2	1	.....	1
September 3, to September 16	2	.....	2
September 16, to September 23	2	.....	2
September 24, to September 30	3	.....	3
October 1, to October 15	1	2	3
October 15, to October 22	1	.....	1
October 22, to October 30	.....	2	2
	23	9	32

COMPARATIVE TABLE.—SHOWING INCREASE OR DECREASE OF CASES
OF GASTRO-INTESTINAL DISORDERS, MAY TO OCTOBER 1893.

	May.	June.	July.	Aug.	Sept.	Oct.
Cholera morbus	4	5	17	23	8	3
Colic	14	31	21	43	11	36
Constipation	28	45	52	50	69	67
Diarrhea	187	161	343	398	385	299
Dysentery	1	2	3	9	4	1
Dyspepsia	..	2	2	.....	.....	8
Enteritis	1	1	2	3	1	....
Indigestion	61	91	183	164	111	183
Intestinal catarrh	5	1	5	3	4	....
Nausea	12	9	37	26	17	16
Typhoid	2	2	4	6	2	1
Vomiting	4	6	17	7	4	..
Disordered stomach	1	..	..	..	..	..
Flatulence	1	..	..	..	..	..
Intestinal cramps	..	..	8	..	..	..
Totals	321	356	689	782	616	609

Out of 988, 32% in May; out of 1,137, 31% in June; out of 1,447, 47% in July;
out of 1,621, 41½% in August; out of 1,641, 39% in September; out of 1,502,
33⅓% in October. Summary: 3,359 cases of diseases of the intestines out
of 8,646 medical cases, 37, 91-100.

TABLE OF DISEASES AND INJURIES TREATED BY THE MEDICAL BUREAU, WORLD'S COLUMBIAN EXPOSITION, CHICAGO, ILL.

	Construction period.	Ex-position period.	Post ex-position period.	Total. (not including post ex-position period.)
Continued and eruptive fevers	569	929		1498
Malarial fevers	136	291		427
Erysipelas	4	2		6
Syphilis	4	5		9
Rheumatism and gout	30	249		279
Diseases of the nervous system	183	1740		1923
Diseases and injuries of the eye	198	469		667
Diseases and injuries of the ear	7	36		43
Diseases and injuries of the nose	39	44		83
Diseases of the circulatory system	13	50		63
Diseases of the respiratory system	280	367		647
Diseases of the digestive system	999	4452		5451
Diseases of the lymphatic system	15	7		22
Diseases of the urinary system	18	65		83
Diseases of the generative system	21	154		175
Diseases of the female breasts	1	6		7
Diseases and injuries of bone and periost.	8	4		12
Diseases and injuries of joints	10	8		18
Diseases and injuries of tendons and bursæ	13	11		24
Diseases and injuries of connective tissues	23	17		40
Diseases and injuries of the skin	299	636		935
Toxic diseases	12	47	Not classified 619.	59
Edema local	3	3		6
Abscess (not classified)	21	34		55
Burns	67	126		193
Scalds	16	18		34
Sunstroke	4	4		8
Exhaustion	57	....		57
Wounds, (incised, contused, lacerated and punctured)	2373	1568		3941
Sprains	230	133		363
Dislocations	84	18		52
Fractures of the skull	14	6		20
Fractures of bones (other than skull)	200	86		286
Concussion of brain and spinal cord	9	9		18
Amputations	9	8		17
				17,521
Totals	5,919	11,602	619	18,140
Re-treated	5,934	6,697	513	13,144
Totals	11,853	18,299	1,132	31,284
Deaths	32	86	1	69
Births	1	* 6	....	7

* (4 premature.)

MORTALITY.—DURING THE PRE-EXPOSITION PERIOD THERE WERE 32 FATAL CASES IN THE FOLLOWING LOCALITIES:

Mines and Mining Building	2	Agricultural Building	1
Manufactures Building	5	Edgemore Bridge Company	2
Electricity Building	6	Pennsylvania Building	1
Transportation Building	2	Landscape	1
Government Building	2	Administration Building	1
Machinery Hall	1	Horticultural Building	1
Illinois Building	1	Esquimaux village	1
Casino Building	1	German village	1

MORTALITY.—DURING THE EXPOSITION PERIOD TO OCTOBER 30, IN-
CLUSIVE, THERE WERE 86 FATAL CASES, AS FOLLOWS:

Terminal Station	2	New York State	1
Wellington Catering Company	1	Esquimaux village	1
Naval Pier	1	Banquet Hall	1
Java village	2	In front of Agricultural Building	1
Midway (Algerian)	1	Employee of Paine at pyrotechnic	
Manufactures	2	display	1
Ice Sledge	2	In front of Mining Building	1
Visitors	4	Dahomey village	1
Cold Storage	18	At Emergency Hospital	1

During the post-Exposition period there was one fatal case, as
follows: Java village, 1; total number of deaths, 69.

The following notice to visitors was distributed
over the grounds and buildings:

WORLD'S COLUMBIAN EXPOSITION—MEDICAL BUREAU.

NOTICE TO VISITORS.

The following stations have been established within the
grounds of the World's Columbian Exposition for the gratui-
tous treatment of those taken sick or injured during the
dedication services of Oct. 21 and 22, 1892.

Emergency Hospital.—At the southeast corner of the Service
Building, near the Sixty-second Street gate.

Substation 1.—At the northwest corner of the Manufactures
and Liberal Arts Building, facing west toward the Wooded
Island.

Substation 2.—At the southeast end of the same building,
east of the Music Stand and facing the Lake.

These stations will be open continuously on October 21
and 22, and will display the Geneva cross. All employees of
the Medical Bureau will display the same emblem.

Invalid chairs belonging to this Bureau and bearing the
Geneva cross will be found in the side aisles of the Manufactures and Liberal Arts Building. An attendant will be
in charge of each chair and will convey patients to substations 1 and 2.

A chair will also be stationed at the landing stairs on the
west bank of the lagoon, east of the Transportation Building.

The Columbian Guards will furnish information as to the
location of the stations of the Medical Bureau.

(Signed.)   JOHN E. OWENS, Medical Director.

www.ingramcontent.com/pod-product-compliance
Lightning Source LLC
Chambersburg PA
CBHW022034190326
41519CB00010B/1706